W9-CFB-369
NEVADA COUNTY LIBRARY - GRASS VALLEY
DEC - 1999

PARROTS

A PORTRAIT OF THE ANIMAL WORLD

BEN SONDER

TODTRI

NEVADA COUNTY LIBRARY - GRASS VALLEY

Copyright © 1997 by Todtri Productions Limited. All rights reserved.
No part of this publication may be reproduced, stored in a retrieval system
or transmitted in any form by any means electronic, mechanical, photocopying
or otherwise, without first obtaining written permission of the copyright owner.

This book was designed and produced by
Todtri Productions Limited
P.O. Box 572
New York, NY 10116-0572
Fax: (212) 695-6988

Printed and bound in Korea

ISBN 1-57717-067-9

Author: Ben Sonder

Publisher: Robert Tod
Editorial Director: Elizabeth Loonan
Book Designer: Mark Weinberg
Senior Editor: Cynthia Sternau
Project Editor: Ann Kirby
Photo Editor: Edward Douglas
Picture Researchers: Laura Wyss, Meiers Tambeau
Production Coordinator: Jay Weiser
Typesetting: Command-O Design

PHOTO CREDITS

Photographer/Page Number

E.R. Degginger 4, 8–9, 17, 29, 30 (top), 33, 35 (top & bottom), 38 (top & bottom), 46 (bottom), 47, 51, 55 (top), 56–57, 61, 64, 66, 67 (top)

Dembinsky Photo Associates
David Hastings 71
Anthony Mercieca 21, 28, 43
Stan Osolinski 67 (bottom), 70 (top & bottom)
Dusty Perin 10

Nature Photographers Ltd.
Robin Bush 44
Kevin Carlson 18 (bottom), 31, 60 (bottom)
Hugh Clark 34
Peter Craig-Cooper 65 (bottom)
R.S. Daniell 46 (top)
Walter Giersperger 58
Michael Gore 65 (top)
Paul Sterry 19
Rick Strange 7

Picture Perfect 45
Steve Bentsen 27
Gerald Cubitt 59 (bottom)
William Folsom 37
Joe McDonald 54

Tom Stack & Associates
Nancy Adams 12 (top), 26 (bottom)
Joe Cancalosi 12 (bottom), 18 (top), 48, 44 (bottom), 60 (top)
Mary Clay 39
W. Perry Conway 13
Kerry T. Givins 6
Barbara von Hoffmann 20 (bottom), 30 (bottom)
Chip & Jill Isenhart 3, 36 (top & bottom)
Brian Parker 11, 68–69
John Shaw 32 (top & bottom)
Larry Tackett (5, 40–41, 62
Roy Toft (16, 20 (top), 22, 26 (top), 50 (top), 53
Dave Watts 49, 59 (top)

Lynn M. Stone
15, 24–25, 32, 50 (bottom), 63

VIREO, Acadamy of Natural Sciences, Philadelphia
C. Munn 14
T.J. Ulrich 23
Doug Wechsler 42 (bottom)

INTRODUCTION

This great green macaw, also known as a Buffon's macaw, wears a cap of downy feathers ending at its cere in a tuft of brilliant red. A native of Central America, it has, unfortunately, been added to the list of endangered birds.

The long, interwoven history of parrots and humans goes back at least as far as the middle of the first millennium B.C., when these birds were first kept in captivity. Before that time, they appeared as wild birds in the folklore, religion, and poetry of ancient cultures. About 3,400 years ago, the oldest surviving piece of Indian literature, the Rigveda, *assigned parrots the early morning role of guardians of the fading moon. Much later in India, they appeared as talking companions—jesters and wits—the prized possessions of princes and noble people. Teaching them to talk became part of the Indian nobleman's mastery of the sensual, as spelled out in the* Kama Sutra, *soon after the beginning of this millennium.*

Parrots arrived in Europe in 327 B.C., when one of Alexander the Great's sailors brought some back from a campaign in India. Soon after, parrots became accessories to the niceties of Greek civilization; by the time of the Roman empire, a parrot that could mimic a few Latin phrases was worth more than a slave. Unfortunately, the parrot's verbal ability did not always act in its favor. In classical times, a diet of parrot tongues was prescribed for those who lacked eloquence and had trouble speaking. During the Middle Ages, the parrot's ability to speak won it admiration from devout Christians. The Vatican accorded talking parrots special status and a closer place to God than nonlinguistic animals. By the late fifteenth century, parrots were associated with more earthly concerns. Many of the explorers of this period were convinced that where there were colonies of parrots there was also gold.

What exactly sets this bird so favorably apart from others that can be kept in captivity and tamed? And why does it occupy such an important place in the literature, art, and music of so many cultures? One of the easiest answers is that the parrot is adept at deceiving us. It will often adapt to a life of shoulder-perching even though it might prefer the wild. It can mimic human ideas and feelings by memorizing phrases from the languages we speak, even if it does not often understand their meanings.

Another reason why the parrot attracts us is that the breathtaking, almost shocking plumage of conflicting colors possessed by many species delights and perplexes the human eye. Such exhibitionism of appearance might very well be criticized in the context of human society, but when it is confined to the animal world, we marvel at its boldness and energy.

Whatever our reasons for valuing parrots, most of our compliments condemn them to a cartoon image. The parrot of the human imagination, confined to its perch, its birdseed, and its doting pet owner, is actually a vastly different creature than the parrot of the forest or mountainside. In our haste to find an imprint of our own habits and feelings elsewhere in nature, we've forced "Polly" into an extremely simplified stereotype. To the average pet owner, she may be little more than a good comedian, something between a complicated toy and a friend, or an annoyance whose ear-splitting early-morning shrieks wakes us too early and makes the baby start screaming. We may claim to fancy parrots, but most of us remain unaware of the variety and scope of this family of more than 300 species with their many physical variations and many complex strategies for survival. Few of us are aware that there are parrots that can last through harsh winters without migrating, winging their way across vast fields of snow. Few of us are aware that there is a parrot species than can construct

This caninde macaw keeps its magnificent blue and gold plumage carefully preened. Natural balance and specially adapted toes make it a superb acrobat that can perform almost any necessary maneuver at the outermost tip of a branch.

The beak of this black-capped lory may look typically parrot-like, but it is actually much weaker than the beaks of seed- and nut-eating parrots. The tongues of lories are covered with erectile tissue ideal for licking viscous liquids.

enormous nests of twigs like condominiums, which can house several families of birds.

Many of us are delighted by the acrobatic antics of parrots, but we rarely consider the amount of agility and coordination required to hang swinging from a branch by one leg while the other grasps food and moves it toward the mouth. And few of us are aware that parrots can eat seeds and nuts too hard or too poisonous for other bird species or that some parrots crave the taste of clay, which protects their stomachs from harsh alkaloids.

If we encountered parrots in nature, most of us would be hard-pressed to explain the reasons for their behavior. But a deeper understanding of the behavior of each species can only deepen our appreciation of this complicated and highly evolved bird. Understanding the characteristics and survival patterns of parrots can help us ensure their existence on this planet. However, in the case of some species, it is too late to understand. The last Carolina parakeets in the world died in a Cincinnati zoo in 1914. They were the only species that North America could claim as indigenous. Giant red parrots, supposedly spotted by Christopher Columbus on Guadeloupe in 1496 are, if they existed at all, gone forever.

The list of extinct parrots is long and lengthening. It is too late for them, and, meanwhile, other endangered species are hanging on to existence by a thread. Two forces against parrot survival continue to wreak havoc. The first is the clearing of tropical forest land for homes or agriculture. In Mexico, Colombia, Ecuador, and Peru, diminishing forest land has put thirteen species of parrots in danger of extinction. The Mauritius parakeets, once numerous, have been reduced to five males and three females, while every attempt is currently being made to save them. The same species had already been exterminated on the nearby island of Reunion by 1800, as a result of deforestation and hunting.

The white-tailed black cockatoo can flourish in all types of woodland. It adapted quickly to the Pinus trees introduced into its habitat, making a delicacy of their seeds.

The second force against parrot survival is directly related to that human stereotyping discussed earlier. Because humans have cast the parrot in the role of entertaining pet, the parrot trade has become big business. The rarer the species, the more people are willing to pay for it, and this despite the fact that a species' rarity is also a signal of its inevitable extinction. Many of the deals made in the parrot trade are illegal, necessitating smuggling techniques that can result in the death of the bird. But since a rare species like the Major Mitchell's cockatoo can bring in as much as thirty thousand dollars, bolder and bolder chances are being taken. Parrots that cannot mate in captivity are being bought and sold without the slightest thought for their future.

A greater understanding of the special needs of parrots in the wild would reverse these trends. Those who admire the parrot need to work to demystify parrot stereotypes and encourage others to see them as animals functioning within specific habitats under certain conditions. The chances for the parrots of the future lie in our willingness to take a closer look at their lives, their needs, and their destinies.

The yellow-headed parrot, a native of the Amazon basin. This bird prefers arid forests, but small groups have adapted to neighboring urban areas.

Following page: **Galahs hunt for food both in trees and on the ground. On the ground they seldom dig for morsels, preferring to eat seeds, sprouts, or insect larvae that are visible on the surface.**

THE PARROT WORLD

Parrot is the name applied to members of a large bird order known as the Psittaciformes. The 328 living species of these birds are located mostly in the southern hemisphere and cover a range of body forms. These body forms are as familiar to us as that of the little pet-shop parakeet, or budgerigar, with its lime-green belly, or as exotic to us as the hawk-headed parrot of South America, with its full-circle crest of maroon and blue feathers surrounding its entire head. Parrots also range widely in size, from the hyacinth macaw, which, counting its tail feathers, can reach a length of over 3 feet (1 meter), to the tiny pygmy parrot of New Guinea, which never grows much beyond 3 inches (7.6 cm).

Body Characteristics

Nearly every parrot, large or small, shares several common characteristics, including a curved, hooked bill and short legs. Parrots also have a projection of flesh, called a cere, at the base of the upper bill into which the nostrils are set. The upper bill of parrots is freely hinged and the lower half is fastened to the skull by ligaments in a way similar to the human jaw. This arrangement gives the bill powerful leverage, so much so that some parrots can crack large nuts that a human would have trouble cracking with a large pliers or hammer. Like most other seed-eating birds, parrots have an ample crop, which is a pouch in their esophagus where hard seeds are softened. They also have a gizzard that grinds the seeds against fine gravel.

Most parrots have thick tongues. The majority of species use this tongue to hold and position food. However, some parrots, such as the lories, have a brush-like tongue, covered with erect papillae that they use to extract nectar and pollen from flowers.

This blue-and-yellow macaw from Brazil exhibits the typical characteristics of its order: a large, hooked bill topped by a fleshy projection into which the nostrils are set. The lower half of its bill is fixed and notched for the positioning of large seeds or nuts.

The vivid colors of this rainbow lorikeet make it appear disturbingly conspicuous. However, seen from above, its green wings and posterior provide perfect camouflage against the foliage where it feeds.

Lories thrive on the delicate, sticky nectar and dusty pollen of flowering plants. They comb the blossoms with their papillae-covered tongues, sometimes stripping the flower of its anther or crushing the blossoms into limp filaments.

Kea have extraordinary survival skills. They make do year round in alpine forests and mountain scrub lands. They use their narrow, curved bills to hunt for rootlets or larvae, even under a cover of snow.

The first and fourth toes of a parrot's legs are turned backwards, allowing the toes to encircle twigs as the parrots climbs from branch to branch. It would be more correct to think of a parrot as having three legs, for the majority use their strong beaks to grasp the branch above or in front of them, hoisting themselves up as mountain climbers would use a grappling hook to move further up a mountain. Some parrots also use their legs for holding food, working their feet in conjunction with the bill and tongue during feeding.

Parrots have fewer feathers than many other birds. However, the feathers that they do have tend to be strong. The parrot family exhibits a bright spectrum of feather colors, the most common of which is green. Vivid reds, yellows, and blues work with the green to serve as camouflage or are flashed at key moments as a means of signaling other parrots or distracting predators. In many cases, the color of a parrot's feather is actually an optical illusion. For example, the deep purple of the hyacinth macaw's feathers is created when its brown feathers scatter light through their millions of tiny air vacuoles. It is the same effect that is produced when particles and droplets in the air are scattered, making the sky appear blue. The sexes of most American and African parrots have nearly identical plumage, but among species in Asia and the Pacific, there is a contrast in plumage between the sexes. This contrast is most marked in the red-sided eclectus of New Guinea. The male of this species is green, whereas the female is mostly red. A few species of parrot are black, gray, or khaki-colored.

Walking on their short legs, parrots look like tipsy waddlers. However, despite their short wings, nearly all parrots are adept and swift flyers. Two exceptions are a ground-dwelling parrot of Australia, which hides in heathlands and swamps, and the kakapo of New Zealand, which is the size of a cat and has evolved to lose most of its power of flight.

The feathers of a blue-and-gold macaw look brown when held against the light. However, dark melanin granules hidden in the feathers can absorb all red light, reflecting back only the blue. Incident light makes the feathers look sky-blue with a touch of turquoise.

These scarlet and red-and-green macaws have congregated to feast on clay. Kaolin in the clay may protect them from the toxins of such noxious seeds as those of the soap-box tree.

Many parrots, such as this chestnut-fronted macaw, are prehensile. Their first and fourth toes provide the same function as the human thumb, allowing them to fully encircle and grip twigs and branches.

This Costa Rican scarlet macaw, resting in a coconut palm, is found in lowland tropical rainforests. There it feeds on palm fruits, figs, berries, and nuts. During the breeding season, it adds variety to its diet by hunting insects and their larvae.

Where Parrots Live

At this point in time, most species of parrots live in South and Central America. Over seventy species live in Brazil alone. Aside from introduced species, none live in North America or Europe. Only a few dozen of the 328 species of parrots live in Africa, India, and South East Asia. One hundred and nine species of parrots live in Australasia, New Zealand, and the Philippines.

The existence of most parrot species in the southern hemisphere cannot be explained by the fact that the majority of parrots need a tropical climate to thrive. At least seventy species of parrots live below the Tropic of Capricorn, where the weather can get as harsh as it can get in Southern Europe and Asia. In the New York and New Jersey area, domesticated monk parakeets that were set free from cages have established successful habitats, despite the long, cold winters.

A possible explanation for the southerly distribution of parrots is that this thirty-million-year old family of birds once did live in the northern hemisphere but were driven south by the advance of the glaciers during the Ice Age. Early parrots may have evolved on the southern supercontinent of Gondwanaland, before tectonic plate shifts separated this land mass into South America, India, and Africa, leaving a large piece behind to become Australia.

Although most parrots are associated with tropical and subtropical climates, there are some startling exceptions. The most northerly naturally existing parrot is in eastern

The historical range of the superb parrot was severely narrowed by the clearing of box woodlands in northern Victoria, Australia. Conservation of the remaining habitats on private lands and the planting of new trees by farmers is now helping to protect it.

The breathtaking king parrot has shocking red body plumage set off by fluorescent turquoise epaulettes. Some find their way to the suburbs of Sydney, Australia, where they can be seen gliding swiftly over clipped suburban lawns.

The king parrot lives in thickly forested Australian ranges, feeding on seeds, berries, and blossoms in the canopy. Like many parrots, it pairs with its mate for life.

Afghanistan, thirty-four degrees north latitude. The most southerly survives in the beech forests of Chile and Argentina, as far as fifty-five degrees south. On Antipodes Island, a New Zealand parakeet can be seen braving frigid winds coming from the Antarctic as it forages for insects among the penguins and seals. Most parrots prefer dense forests, but that does not mean that some species cannot be found living above altitudes of 11,000 feet (3,355 meters), in the mountains of Papua New Guinea or Tibet. In the New Zealand Alps, a parrot known as the kea can survive many feet above the snow line. Some parrots live near oceans. The budgerigar, which we usually refer to as the common parakeet, can survive in the most arid parts of central Australia.

A Life in the Trees

For the majority of parrot species, the ideal habitat is high up in the trees of a lush tropical forest. Most parrots are "squatters." Instead of building their own nests, they nest in the hollows of trees that have been struck by lightning and eaten out by rot; or else the hollows were dug out by termites. When a parrot invades termites' space, these insects maintain their privacy by building a wall to transform the cavity into a "two-room" space.

Although parrots do not tend to create their own living spaces, many use their strong bills to "redecorate" them by scraping the sides into a suitable compartment. Others, such as some lovebirds, carry leaves and bark to line their nesting cavities. Still other species of parrots move into ready-made nests, driving the previous occupants out. There are a couple of dramatic exceptions to these patterns. The Argentine monk parakeets start from scratch, using sticks to build huge communal nests in the trees. The nests house several families, and each has its own entrance. The nearly flightless Australian ground parrot nests right on the earth, hoping to remain hidden within a tuft of grass.

While the majority of parrots don't build their own nests, they are adept at making use of pre-built structures. This South American blue-winged parrotlet has borrowed the mud-and-grass nest of another bird, the lesser hornero.

In Australia, where there are gum trees—especially eucalyptus—there are likely to be rainbow lorikeets. A flock can strip a flowering tree of all its nectar in a matter of minutes.

At Corcovado National Park in Costa Rica, this scarlet macaw is devouring a beach almond. Even without the help of its prehensile toes, it rarely loses a morsel of food that has been split and crushed by the fine-tuned movements of its chisel-sharp bill.

Outside of their nests, parrots of the forest reign in the topmost levels of the trees. Their expert grip on twigs and branches gives them an agility that reminds one of monkeys. Because their legs are so short, their center of gravity is low, preventing loss of balance. Using their yoke-toed feet and their widely gaping bills, they can pull themselves quickly from branch to branch. Some can hang upside down from their perch like acrobats from a trapeze or, attached only by one leg to a branch, use the other to pluck a seed pod and bring it to their open bill.

The solidity and leverage of the parrot's bill is an important asset to survival in the trees. The bills of some parrots, such as the palm cockatoo, can crush hugs palm nuts with a minimum of effort. In the Amazonian jungle, macaws make short work of the rock-hard shells of Brazil nuts, cracking through to the rich, white meat inside in one swift movement. But strength in opening nutshells must be combined with precision. When a parrot opens a nut, whether it is the size of a sunflower seed or a Brazil nut, the parrot's tongue first rotates it and examines it, taking note of its shape and its seams. It places the nut in an ideal position against the notches in its upper mandible before piercing it with its lower mandible.

Because parrots tend to eat seeds as well as pods, fruit, and some leaves, they are plant murderers. Other animals who eat fruit and who lack the equipment to crush seeds carry them in their intestinal systems and deposit them unharmed in other locations. These animals are fruit eaters, but they are also seed scatterers. But parrots, whether they live on grains, grass, or large nuts, tend to destroy seeds. The parrots grind seeds into a nourishing pulp inside their gizzards, absorbing the nutrients into their bloodstream.

Nature has provided many plants with protection against seed predators. Either the seed has such a hard shell that the predators cannot crush it, or it is composed of bitter

This red-fronted parrot has found a home in a dead tree trunk in Texas. Rot or another animal may have created the hollow, but the red-fronted will remodel it to its liking.

A flock of little corellas takes a break on a barren tree from their search for seeds and berries. In North and Northeast Australia, they frequent open savanna lands and eucalyptus forests.

alkaloids and tannins that will disgust or even poison an animal. But parrots have a remarkable resistance both to tough seed pods and to those containing tannins and alkaloids. They can even eat unripe seeds, when the concentration of these chemicals is at its highest. In some cases groups of parrots can divest an entire tree of all of its fruit and seeds in a matter of a couple of weeks.

Nectar-eating parrots, found in Australia, such as lories and lorikeets, have less powerful beaks than their nut-eating cousins. The nectar eaters use the tiny tufts covering their tongues like miniature tentacles to coax the nectar and pollen of flowers into their bills. They seek out trees like the eucalyptus at a time when their flowers are producing thick flows of nectar. The lories and lorikeets descend on the canopy of the trees, filtering the rich flow of nectar into their bills and sometimes destroying the flower by chewing the anthers in order to eat the pollen.

These regent parrots have learned to survive cooperatively. They have an entire repertoire of signals used to coordinate their activities.

Parrot Society and Breeding

Most parrots survive cooperatively, using a complex system of group solidarity, signaling, and coupling to avoid predators, to feed, and to raise families. It is not uncommon to see entire flocks of parrots wheeling through the sky or settling like a blanket of feathers over land. The whirling mass confuses predators. There is safety in numbers, for it means that many pairs of eyes are constantly scanning for danger.

In order to keep together, parrots imitate each other. When a parrot sees other members of its group feeding, it begins to eat. If the others are moving toward the edge of a pond for a drink or a bath, the parrot takes notice and follows. A parrot's brightest feathers are usually concealed within its tail, rump, or underwings. As it springs into flight, the colors are flashed. These sudden, startling color signals are meant for other members of the flock, who take them as a signal to get going and follow.

Parrots have a whole repertoire of gestures meant to intimidate enemies. They may point their beaks at an aggressor, suddenly raise and lower their wings, or hunch forward. Few

Following page: Many parrots spend most of their lives at the very tops of trees. This blue-and-gold macaw is at home at heights of up to 130 feet (39.7 meters).

When one of these scarlet macaws took flight, the flash of its under-feathers signaled its companions to follow. Their pacing is perfect, as those in tow watch for further directional cues from the leader.

parrots are eager for an actual confrontation. The trick is to slowly escalate the intimidating signals in the hope that they will be enough to ward off a real struggle. Some, such as the Madagascar lovebird, will lock beaks with another bird in a kind of duel, but this is often a ritualized form of wrestling designed to establish dominance without resorting to real injury. Parrots who find themselves in close proximity to a parrot who is not their mate will often turn the situation into a friendly preening session, offering their raised head feathers to the beak of the other parrot for extended periods of grooming.

Two scarlet macaws show their solidarity by interlocking toes and touching bills. Later they may engage in mutual preening. Many parrots have such a strong need for each other's company that they take to the point of lifetime bonding.

Many parrots establish permanent bonds with their mates. Perhaps because many species of parrots do not need regularly to attract new mates or compete with rivals, they exhibit little sexual dimorphism. The male and female tend to resemble one another. The couple stays close to each other during feeding and sleeps together in their nest. When the time for breeding comes around, the birds may already be very familiar with each other. The cock may indicate his intentions by a series of courtship gestures that include bowing, strutting, head nods, and flicks of his wings. These movements tend to show off some of his most attractive color patterns. Male palm cockatoos do their wooing from a perch, hanging upside down with open wings and spread tail, their cheeks blushing red. The female responds by crouching with wings and tail raised so that she can be mounted. When she has had enough of the experience, she threatens the male by squawking over her shoulder.

When the female is about to lay eggs, the male may supply her with extra nourishment. He serves food to her from his crop, where it has been partly digested. Most parrots' eggs are white. They are relatively small, especially among larger species. From these small eggs will emerge puny offspring. In the case of South American parrots, the babies are born featherless and blind, without yet having an external ear opening. Their heads are topped by a bulge of skin from which their crests will one day emerge.

The immature infant birds need twice digested food to survive. It has been passed partially digested from the crop of the male to the hen. Then she has stored it in her own crop to digest it more thoroughly. The hen delivers the food to her babies by pumping into their gullets. As the baby birds grow larger, both parents search for raw food that the youngsters can digest and offer it directly to them. In time the young will be ready to eat all of the foods that their parents eat, including tough-walled seeds. Then they will leave the nest to do their own foraging.

These green parakeets have found safety in numbers. From a distance, they look like uniform clumps of foliage to most predators.

THE PARROTS OF THE AMERICAS

What we tend to think of as the "typical" parrot is a sampling of various species that make up the subfamily Psittacinae. Every single species native to the Americas fits the common description of a typical parrot. In addition, the greatest portion of the total number of species of parrots—two thirds, which is about 210 kinds—all live in South and Central America.

Unlike parrots in other regions of the world, the parrots of the Americas tend to be very homogenous, making it easier to forge generalizations about these species. This does not mean, however, that the vast parrot population of the Americas has been fully understood, described, or even discovered. European explorers first brought back descriptions of New World parrots over 500 years ago, during the Age of Discovery. In fact, Columbus is supposed to have been led southward in the Caribbean by a flock of parrots, which kept him from continuing on to the American continent. Nevertheless, as recently as 1980, a new species of parrot was being discovered and catalogued in Central America.

Amazons

In the last hundred years or so, some of these green, short-winged parrots from Latin America have become cultural clichés. Sailors and traders brought them back from the New World during the late nineteenth century. Those birds that survived often ended up as the pets of inn-keepers, seaside bar owners, or barbers. Since Amazons are the best talkers of the New World parrots, they learned a colorful language from these rough-and-tumble companions. Stories and images of cursing, shrieking green parrots on the shoulders of pirates and sailors soon became part of the common mythology. One of the most well-known and popular of the Amazons, the double yellow-headed Amazon, soon became a staple of the pet bird trade.

In reality, the Amazons are more diverse and more complicated than legend would have it. There are about twenty-seven species that range from the South American continent to northern Mexico. Most Amazons are stockily built and have square

The double yellow-headed Amazon is a favorite of the bird trade. Humans seek them out for their intelligence, colorful plumage, and talkative personalities. They are agreeable birds, unless confronted by another male, whom they are likely to challenge.

This green-cheeked Amazon is sometimes called a Mexican redhead because of its crown of red feathers. In Mexico, the wild species must compete with the ornate hawk-eagle, which feeds on parrots. Green-cheeked males must contend with this competitor, but they often tolerate no other species of their own within their territory.

Yellow-naped parrots frolic in a lush Costa Rican forest. They are the most symmetrically shaped of all the Amazons. To many, their low, inflected voices sound almost human.

tails. The species vary in size, from the 10-inch (25.4 centimeters) white-fronted Amazon to the rare 18-inch (45.72 centimeters) imperial Amazon. Their feathers vary from darkish blue to green to yellowish green. Some have head feathers of red, blue, lavender, or white. However, compared to some other birds, they seem drab at first—that is, until they flare into display to impress a mate or intimidate another bird. Then their fanned tails reveal vivid markings; their head feathers raise and they strut with stiff legs. When the wings are raised from the body, lower wing feathers of more dazzling colors are revealed.

A few species of Amazons exhibit sexual differences. For example, the white-fronted Amazon male has red wing marks and red to the rear of his eyes. A female has no red wing feathers and red-brown irises. In general, Amazons are known for their aggression during breeding periods. When it comes to defending a mate or offspring they can be very menacing. However, different species of

The white-fronted parrot, one of the most abundant and smallest of the Amazons, rarely exceeds 10 inches (3.9 centimeters) in length. It ranges throughout parts of Central America. An assembly of large flocks can produce a deafening racket.

Amazons exhibit different temperaments, which can sometimes be altered by humans. The green-cheeked Amazons were tamed by residents of northeastern Mexico before being exported as pets; these birds easily tolerate humans, but in the wild, they will engage in a shrieking match to drive other Amazon species out of their territory. The red-lored Amazons that frequent humid tropical forests in Mexico, Guatemala, Honduras, and the Bay islands, tend to avoid aggressors. Their strategy is to hide behind foliage until the unwelcome guest disappears.

Amazons vary in their ability to talk and in the quality of their speaking voices. The yellow-naped Amazon is probably the most articulate of all of these birds. Its voice is low and full of intelligent inflection. It also has a pleasant singing voice. Blue fronts are also excellent talkers.

Amazons were once prevalent in the Caribbean, but some species are extinct and others are seriously endangered. Some were driven out by settlers who cleared the land, and others were eaten by European immigrants and their slaves. Of the eleven species that existed in the Caribbean, there are now only nine. At present, the large Amazons that used to live in the Lesser Antilles are few and far between. Smaller green species are more common, except for the emerald green Puerto Rican Amazon, whose last refuge is in the Luquillo Mountains on that island. Millions of dollars have been spent in efforts to preserve this species, but only about seventy remain on the island.

Larger Amazons live quite a long time—sometimes longer than 100 years—but they mature very slowly. They also tend to produce few offspring. These facts make them vulnerable as a species. In the Dominican Republic, only about fifty imperial Amazons remain, the rest having been wiped out by the building of plantations and by hurricanes. In other parts of the Caribbean, the situation for species of Amazons is similar. The St. Vincent Amazon, with its orange, blue, and yellow-striped tail, is the native bird of that island. But this has not saved it from hurricanes, cyclones, volcanic eruptions and the vandalism of its nests. Only about 420 of these parrots still live in a small forest in the center of the island. Their future is uncertain.

The Cayman parrot, national bird of the Cayman Islands, has become one of the world's rarest Amazon parrots. Caymanians kept them as pets in the past, but new laws make it illegal to take the birds from the wild for any reason.

Macaws

Gaudy and grand, these long-tailed, strong-billed parrots of the tropics are considered the largest parrots only because they possess very long tails. In terms of weight they actually take second place to the cockatoos. They also vary in size. The hyacinth macaw, at about 40 inches (1 meter) is the largest, and the red-shouldered macaw, at about 12 inches (30.5 centimeters), is the smallest. These birds get their names from the fruit they feed upon, that of the macaw palm, which is found in the Amazon region. They live in woodland areas, from Mexico to Argentina. The green macaws are found from Panama to Paraguay; scarlet macaws are found further north in Mexico.

Because of a large demand for them, macaws find their way into pet stores and zoos both legally and illegally. Most of these animals in captivity belong to the blue-and-gold or scarlet species, whose numbers are still fairly plentiful. This is partly due to successful efforts at breeding in captivity, despite the fact that attempts do not always meet with success. Macaws have one of the longest nesting periods of any parrot species. They tend to lay a couple of small eggs, which

Called the "king" of parrots, the hyacinth macaw, with its cobalt-blue plumage, has become somewhat of a rarity in the wild. There are less than 3,000 left in Brazil. Hyacinth macaws are slow breeders, a fact that makes their declining numbers even more worrisome.

Deep in a Brazilian rainforest this gentle scarlet macaw finds a plentiful supply of palm fruits, figs, and berries to live on. Note its stark white, unfeatured face.

A scarlet macaw on a palm frond in Costa Rica. These birds eat between forty and fifty different kinds of nuts. Within seconds their powerful bill can remove the seeds from thick pods whose walls have the strength of forged steel.

A closer look at a green-winged macaw reveals its horn-colored upper mandible and black lower mandible. It finds its home in the tropical lowlands at the borders of the jungle.

Blue-and-gold macaws are not difficult to identify. They have blue upper parts and a bright yellow breast and body. Their green forehead mutates to blue as it reaches the crown. Stripes of black feathers mark its otherwise naked face.

Also called "Buffon's macaw," this great green macaw used to be common on both sides of the Panama Canal. Its numbers have diminished with its loss of habitat. The birds rarely form large flocks, preferring small groups of five or six relatives, confined to small areas where food is available.

An orange-chinned parakeet. Only close inspection would reveal the bright orange chinspot. It builds its nest in treeholes left behind by woodpeckers or termites.

require a long incubation period. Their chicks, which are born blind and helpless, stay with the parents for as long as three and a half months.

Some macaws have been severely endangered by unscrupulous traders. The hyacinth macaw lives in the riverside areas of savanna forests in Bolivia and Brazil, where it feeds on palm nuts. Its bill is huge, with a long upper mandible that curves downward. But a reduction in palm trees and incessant poaching by traders have made the bird vulnerable and rare in these countries. Nevertheless, some pet owners are willing to pay thousands of dollars for these birds. They love their cobalt-blue plumage, ringed around the lower mandible by yellow. Another blue macaw, known as a Spix's macaw, has been reduced to a single individual. It lives in the bush in the sun-baked village of Curaccedila of northeast Brazil. When this bird dies, only sixteen macaw species will remain. Several West Indian species are already extinct.

The peach-fronted conure, a member of the largest genus of parakeets in the tropical Americas. Surviving on a diet of fruits, seeds, and berries, it can be found in South American savanna or open forest and up to a height of about 3,500 feet (1,068 meters).

Conures and Parakeets

"Parakeet" is the common name applied to many species of the parrot family in the Americas, Africa, Asia, and the Pacific. However, you won't find much agreement even in the English-speaking world over the use of the term. A great many New World parakeets belong to a large genus found throughout the tropical Americas that consists of nineteen species that are known generally as "conures." One of the most familiar conures is the sun conure, a yellow or orange bird ringed in gray or white with wing coverts that are spotted green. Other South American conures belong to the genus *Pyrrhura*. These birds are small to medium in size and have collars with feathers edged in white. They have featherless ceres and long, tapered tails that are usually maroon-colored.

It's hard to determine the sex of an adult conure, as there is little difference in appearance between the sexes. They survive in the savanna and palm groves of Central South America, where they hunt for seeds, nuts, and fruits. Some, such as the Nanday conure, have learned to live near human settlements, breeding in hollow fence posts and eating cultivated crops. Such a talent for adaptation to human settlement worked tragically against a close cousin of these birds. It is known as the Carolina parakeet, and it became extinct in the southern United States partly because it was considered a pest to fruit-tree growers.

One very well-known species of parakeets with a genus all its own is the Quaker parakeet, also called the monk parakeet, a native to Brazil, Uruguay, Bolivia, and Argentina. This green, gray-cheeked bird has a bluish-tinged forehead. Its feathers are edged with white. The Quaker parakeet survives in dry savanna woodlands, as high up as 100 feet above sea level. It eats seeds, fruits, berries, insects, and insect larva. It is unique among parrots in the ways that it builds its nests. Quakers breed in large colonies. They cooperatively build hugs nests from twigs that can weigh more than 2,000 pounds. These huge "condominiums" house whole populations of Quakers, providing them with a home even when trees become scarce. Quakers are inexpensive and immensely popular as pets. Thousands live in the United States and Europe, and some now make their homes in city parks or suburban woodland, after having escaped from cages.

About a foot in length, this jendaya conure is a loud bird, a distressing characteristic for some who try to keep it as a pet. One way to recognize a jendaya is by the white skin patch around the eye.

White-bellied caiques differ from the black-headed variety primarily in their lack of black pigment. However, baby white-bellied caiques may sport a few black feathers on their crowns.

Following page: A quartet of endangered blue-and-yellow macaws share a perch. Pair bondings can be intense among this species. In captivity bonding may occur with a human mate or even another pet.

These preening sun conures live in open forestland in northern Brazil. It's almost impossible to determine sex by their appearance, although females tend to have a rounder and smaller head than the male.

A green-winged macaw takes flight. Its red plumage and horn-colored upper mandible sometimes make people confuse it with the scarlet macaw.

The Saint Lucia parrot's habitat spans a mere 20 square miles (51.8 square kilometers) of forestland at the center of the island. Despite the dense protection of the forest canopy, the entire population dropped to fewer than 100 by 1971. Conservation efforts and new forest preserves brought the number back to about 300 by 1995.

Caiques are medium-sized parrots, identifiable by their white breasts or bellies. Black-headed caiques like this one will gather in small noisy parties or flocks of up to thirty birds.

THE PARROTS OF THE PACIFIC

The parrots of this region of the world are astonishingly diverse. Among the 109 species that live in Australasia, New Zealand, and the Philippines, there are unexpected and fascinating variations. In Australia, the ingenuity of parrot design rivals that of the pouched animals of that vast island. In New Zealand there are even more curious parrot species, with Maori names that begin with the letter "k": the kea, the kakapo, and the kaka.

Australian parrots and plants have had over thirty million years to adapt to each other. More than a thousand Australian plants produce flowers rich in nectar and pollen, and among the insects and other birds that take advantage of this food source, there are several species of nectar-eating parrots. These parrots lack the characteristically powerful bills of other species, but their tufted tongues can collect fluid and tiny particles from the delicate blossoms. Those parrots that are seed eaters are no less adapted to the plant life of this part of the world. Some species can crack the powerful seed capsules of gum trees. Others have beaks like curved spikes that can withdraw the seeds from their capsules like a siphon. Some Pacific parrots take the role of woodpeckers, digging out grubs from the rotten bark of tree trunks. They may also chisel the rotting cavity of a tree trunk for months, until it is just the right shape and size for nesting.

Lories and Lorikeets

There are about sixty species of this bird, belonging to eleven genera. All of them are found naturally in Australasia, New Guinea, and the many islands of the Pacific. They are small to medium-sized birds, ranging from about a foot to almost a foot and a half long. The smaller species are usually called "lorikeets."

Rainbow lorikeets are masters of display. Making use of the multicolored opportunities provided by their plumage, they draw on a repertoire of more than thirty variations of head movements, hops, wing flappings, tail waggings, and struts when they go courting.

The kaka is a fairly common sight on the offshore islands of Kapiti and Steward in New Zealand. Its success seems to depend upon how much nectar it can consume before the breeding season starts.

Scaly-breasted lorikeets are found in northeastern Australia, ranging from Cooktown, north of Queensland, south to the Illawarra district of New South Wales. Their call resembles that of the rainbow lory, but it is higher pitched.

A red lory from Indonesia. The young typically show black at the tips of their feathers, producing a mottled effect.

The "box-cutter" beak of this eastern rosella is highly efficient at tearing open fruits. The bird favors the forest habitats of southeastern Australia, where it usually lives in small flocks.

Lories and lorikeets are nectar feeders. But whereas other nectar-eating birds (such as the hummingbird) suck the sweet fluid out of blossoms by means of a tubelike tongue, lories and lorikeets usually crush the blossom, then lick up the honey with their brush tongue. Sometimes, their feeding becomes so eager that their necks are coated with the sticky nectar and dusty pollen. If it ferments before it can be preened away, it may intoxicate them, and they must wait until they become sober to fly.

One of the most numerous of this kind of bird are the rainbow lorikeets, a genus of colorful small to medium-sized birds who live in the Philippines, on Borneo, along the northern and eastern coastal forests of Australia and Tasmania, and elsewhere in the Pacific. Two other, related genera of lories are the black lories and the red lories, both of which are medium sized with a rounded tail.

Some lories and lorikeets possess ingenious strategies for surviving in the canopies of trees. For example, the celebes hanging lorikeet can rest by hanging upside down from a tree branch like a bat. Lories and lorikeets also have devised ingenious ways of communication. The rainbow lorikeet has thirty different postures designed to intimidate rivals and express ascendancy. It uses these gestures in different sequences all the time, an effective way to startle and drive away a threatening neighbor.

Cockatoos and Cockatiels

Cockatoos are not "true" parrots. They belong to the family Cactuidae, rather than the family Psittacidae, which is considered the true family of parrots. There are about sixteen species of these birds. Most have strongly curved, sometimes massive, bills, and short, blunt tails. They can be black, gray, pink, or white, with yellow or red crests that are always at least partially erect. These dramatic crests are fully raised in a variety of situations, including fright, landing after flight, and during sexual excitement. A cockatoo will also raise its crest to invite another bird to groom it.

Sexual differences among pairs of cockatoos exist only in the gang-gang and the blacks. In the blacks the hens have spotted plumage. The red-tailed black females have no crest. Both the galah, the white, and the sulfur-crested cockatoos can be sexually identified by eye color: The cocks have black eyes, and the eyes of the hens are brown.

All cockatoos enjoy eating grubs, which they tear from rotten wood after stripping away the bark. Big black cockatoos have been known to invade plantations and fasten themselves to the tree trunks, listening for the noise of larva as they strip the bark away, damaging trees a great deal in the process.

One of the most well-known of the cockatoos is the sulfur-crested cockatoo, a fairly large cockatoo with snow-white plumes and ear-covers of pale yellow. It has a large, sulfur-colored crest that reminds one of a sunburst. Most sulfur-crested cockatoos nest in the cavities of large trees. They eat berries, nuts, flowers, buds, roots, insects, and larvae.

Notorious bathers, galahs of the Australian woodlands have been glimpsed hanging upside down during a rainstorm, letting the water trickle through their spread wing feathers. They are not very territorial and will nest near other birds of their species without many disruptive occurrences.

Upon seeing its own reflection, this pink cockatoo reacted by rearing its magnificent head plumage. Erect crests are a sign of strong arousal, used to communicate a variety of signals, from sexual stimulation to fear.

This fiercely territorial Leadbeater's cockatoo, also known as Major Mitchell's cockatoo, is from the arid parts of western and southern Australia. Concealed in its cotton-candy-colored plumage are crest feathers of a startling red, which it may suddenly flash in conjunction with multiple nods of its head.

Two Moluccan cockatoos bond. Their backward-curving crests of coral and yellow have earned them the name "red-vented cockatoo." When the vent is raised, plumage on the cheek also puffs out, swelling the face impressively.

The Major Mitchell's, or Leadbeater's, cockatoo has white plumage and a small, forward-bent crest of deep pink, fringed in white. These birds can be seen in the bush of western Australia, but some have been spotted perched on the fence posts of wheat fields. As the bush becomes settled, the Major Mitchell's cockatoo is becoming endangered. Because these cockatoos have a belligerent disposition toward others of their kind, they need space to survive. Only about twenty-five pairs can survive on an area of land of about 300 square miles, and they cannot breed easily in captivity.

Salmon-crested, or Moluccan cockatoos, are from the southern Moluccas in Indonesia. In general, they have salmon-pink plumage with a darker crest of the same color. Their eyes are black with a blue-tinted ring. In the 1980s, they were placed on the endangered species list, yet some people still keep them as pets. In the wild, these birds survive on

The cinnamon pearl cockatiel of Australia, a common pet parrot. An astounding variety of color variations have been engineered by breeders. The feathers of this bird are dusty, containing a powder that keeps the wings waterproof.

This brilliant crimson rosella possesses sophisticated survival skills. It is adept at locating dead eucalyptus trunks for nesting, spending a considerable time chipping away a hole as an entrance. It can feed even in the snow-blanketed areas of Australia's Great Divide.

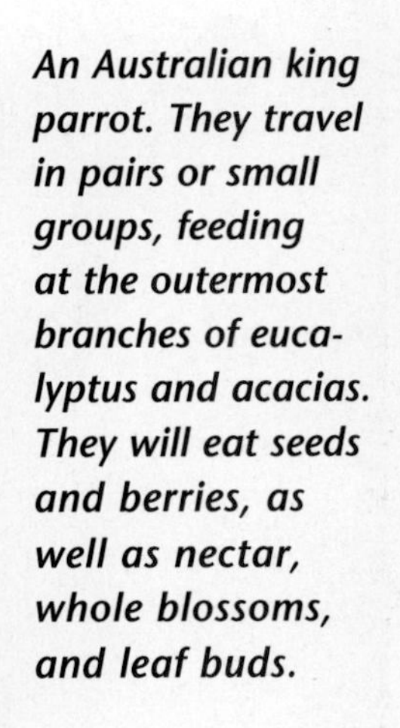

An Australian king parrot. They travel in pairs or small groups, feeding at the outermost branches of eucalyptus and acacias. They will eat seeds and berries, as well as nectar, whole blossoms, and leaf buds.

This scarlet-headed gang-gang, with its sparse crest of curly feathers, can survive in the southern mountains of Australia's Great Divide, which is often covered by a blanket of snow.

berries, fruits, and nuts, as well as insects and insect larvae.

Cockatiels are so common as pets that they are beginning to rival parakeets in popularity. Breeding in captivity has given them hundreds of color variations. In the wild, they can survive on grass seeds in most kinds of open country, and the young are ready to leave the nest in four to seven weeks. The standard cockatiel is gray with white patches on the wings. The face tends to be yellowish with a bright orange patch over the ear canal. A cockatiel waterproofs itself by excreting a water repellent dust that coats its feathers. This dust comes off on one's finger if the bird is touched.

Keas, Pygmy Parrots, and Budgerigars

The kea is a parrot with a perhaps undeserved sinister reputation. It is an olive-colored bird about the size of a raven with red and yellow wing markings. Most kea live among the sparse vegetation of New Zealand's South Island. They can survive at altitudes as high as 6,000 feet (1,830 meters) and build their nests in the rocks. Keas are omnivores. Like lories, they have brushed tongues that can

This typically dark-olive-colored kea is fearlessly curious. Sometimes its search for seeds, rootlets, and insects leads it to rubbish dumps, where it will even devour pieces of rubber tire. Because a few have attacked sheep by night, they were once considered a nuisance. Now the New Zealand government protects them.

Kaka belong to the same family of parrots as the kea. However, unlike the kea, the kaka shies away from human civilization. It prefers lower forestlands far away from human settlements.

The little corella survives on seeds, nuts, flowers, and insects in the open savanna lands and eucalyptus forests of north and northwest Australia. Near the coast it favors mangroves as a habitat.

Following page: The underwing coverts and undertail of this lesser sulphur-crested cockatoo match its yellow crest. It is smaller than the sulphur-crested cockatoo and has a very loud call.

sift nectar from blossoms. But during the winter, hungry keas have learned to congregate near sheep ranches, where they feast on the carcasses or heads of dead sheep. Occasionally, a kea will attack a living sheep, plunging its beak into the sheep's back until it reaches the kidneys. This has made keas the enemies of sheep ranchers, who have killed thousands of them over the years.

There are six species of pygmy parrots, who are—as their name suggests—the midgets of the parrot world. These tiny birds live in the forests of New Guinea, New Britain, and New Ireland. Their stiff tails ending in spiny shafts make them resemble woodpeckers. Their long toes and claws help them keep their balance on the trunks of trees. How they feed is a mystery, but researchers are beginning to suspect that they are fungus eaters, eating the lichen, as well as an occasional insect, from the barks of trees.

For most people, budgerigars may be the least mysterious of parrots. They do well in captivity, and are often referred to by their owners as "parakeets." In the wild they congregate in large flocks, migrating through the

Two courting budgerigars, the most common pet birds in the world, often referred to simply as "parakeets" in the United States. In the wild, huge flocks of these birds migrate long distances in the Australian outback.

The highly threatened orange-bellied parrot is now on the IVCN Red List of Threatened Animals. It breeds only in the wilderness in Tasmania and Australia.

One of the few remaining of the mysterious kakapo, also known as the owl-faced parrot. A native of New Zealand, it is the only nocturnal, flightless parrot in the world.

Australian outback for long distances in search of seeds, greens, carrots, and apples. Budgerigars are often olive green with a yellow crown. Their forehead feathers are edged in black. These small birds are surprisingly good talkers; some can learn to speak over 200 different words.

Ground-living Parrots

Isolation has produced species of incredible originality in Australia and New Zealand. In a few forests of New Zealand, lives the owl parrot, or kakapo, a flightless nocturnal parrot that has the size and appearance of an owl. It is the heaviest of the parrots and can reach a weight of almost 7 pounds (2.6 kilograms). Although the owl parrot can't fly, it can climb trees, where it sometimes hops from limb to limb. Most of the time, it stays on the ground, feeding on small roots, leaves, and soft twigs and nesting under rocks or tree roots.

Kakapo were once common throughout all the islands of the New Zealand Archipelago, but Polynesian settlers in the eighth century

This mallee ringneck parrot makes its home in the mallee scrub of Australia. Only 200 years ago, one-fifth of Australia was covered with mallee trees. Today that habitat is all but gone.

soon began hunting them for food and skins. It was fairly easy to find a kakapo in the grasses, since at night it produces a repetitive, booming call. In the late nineteenth century, European settlers, cats, and dogs killed a large percentage of those kakapo that had been left over from the Polynesian exploitation. By the 1980s, only a few dozen of these birds existed on two islands.

Australia has two kinds of parrots that live on the ground. One is known simply as the ground parrot. It's a green bird about the size of a raven that makes its home in the swamps of the east coast. The ground parrot is a shy nocturnal bird that scurries about on legs that are much longer than most parrots'. Closely related to this parrot is the night parrot, which lives in the arid center of the Australian continent. The night parrot's brown- and yellow-spotted feathers offer it perfect camouflage in the stiff grasses of that region. Few people have ever seen this bird, which only ventures out at night.

An Australian blue-winged parrot favors monogamous solitary pairing during its breeding season, from October to February. Partners choose tree hollows or stumps and logs to breed and raise their young.

The rosella, noted for its plush, ermine-like plumage. Like many other broad-tailed Australian birds, rosellas abhor body contact, reserving it for fighting or mating.

THE PARROTS OF AFRICA AND ASIA

Only thirty-four kinds of parrots live in Africa, India, and Southeast Asia. However, despite the relatively small number of species, the parrots of these regions exhibit an interesting diversity. Some have short tails and resemble Amazons in size and structure. Others look like tiny parakeets.

There are parrots of Africa and Asia that make their home in lush tropical forests and feed on oily palm nuts; there are also those that live at the fringes of human settlements, taking advantage of ripening grain in the fields. Some African and Asian parrots are richly colored, whereas others approach drabness. The vasa parrot of Madagascar, for example, may be the drabbest parrot on the planet. Its dull brown feathers make it look almost like a dirty city pigeon.

The Gray Parrot

The gray parrot, found all over tropical Africa, is one of the parrots of this region that resembles an Amazon. It is a gray bird that grows to about a foot in length and has a short red tail. There are two subspecies of this bird, the Congo African gray and the timneh African gray. The Congo African gray is the more common species. Timnehs are confined to smaller areas of Africa, within Liberia and the Ivory Coast. African grays feeds on many kinds of seeds, fruits, and nuts; but they are especially drawn to the tropical oil palm. Those that live near human settlements have been known to raid cultivated crops.

The African gray's ability to imitate the human voice has made it an immensely popular pet. For this reason, tens of thousands are exported from Africa every year. Some are exported from the Cameroons, where the population of grays is so plentiful that the

The African gray, the most accomplished linguist of the parrot world. Its deductive skills—exhibited by counting or recognizing and naming objects—rival those of the chimpanzee.

Large flocks of red lories fly between islands in the southern Moluccas in search of nectar, fruit, and pollen. Occasionally an unpleasant confrontation results in a session of hostile posturing.

people who live there tend to regard them as pests. But others are exported from Mali and Togo, despite the fact that African grays are becoming very scarce in these countries.

Recent experiments have shown that the African gray may not just be a skillful mimic. Some African grays may have cognitive abilities equal to those of porpoises and primates. At the University of Arizona, an African gray named Alex was successfully trained to count up to six and to recognize or name about a hundred different objects. He can also identify their color, texture, and shape.

A Brown-headed poicephalus poised at the entrance to its tree-trunk nest. They favor moist woodland, where they profit from a variety of seeds, nuts, and berries. Some have been spotted eating nectar as well.

The Poicephalus Parrots

Poicephalus parrots live on the African continent in a vast range stretching from the equator to the cape. "Poicephalus" means "green head," but the name is inaccurate since no species of poicephalus has green plumage on its head. These medium to small-sized birds are usually brown. Their tails are squarish and their bills are large. Some of the species of this bird, such as the Senegal, are common as pets.

There are nine major species of poicephalus parrots that divide into about twenty-five subspecies. Some, such as the Senegal, live in moist woodlands or at the edges of savanna. They eat seeds, fruit, and grain. Many farmers think of them as pests because they raid fields of maize and millet or steal peanuts.

Another species of poicephalus, called the jardine parrot, is larger than most of the other species. Jardines resemble small Amazons. They live in an area spanning Central Africa, from Liberia to Tanzania. These birds congregate in lowland rainforests or mountain cloud forests. They feed on fruits, wild olives, and seeds. Sometimes they stray to the forest

Rosy-faced lovebirds are superbly equipped for nest-building in the wild. They forage for long grasses and wood fibers, which they pack into their rump feathers, to be used to line their nests.

The peach-faced lovebird, known for its frontal band of deep rose-red and its paler rose-red throat. These birds can be spotted in groups of ten in the dry country of South Angola.

Lovebirds are known for their strong pair bonding and frequent allopreening. This black-cheeked variety are found only in two river valleys, one in southwestern Zambia and the other in the Victoria Falls area of Zimbabwe.

edge during the day in their search for food, but by nightfall they are back in the forest canopy.

Lovebirds

Although people tend to think of lovebirds as any small species of parrots that remain close to each other in pairs when caged, the name really refers only to eight specific species belonging to the genus *Agapornis* and found in Africa and Madagascar. Lovebirds are small parrots, rarely larger than half a foot. They have large heads and short tails; and their bodies are green, with red, yellow, black, or gray head, neck, and breast markings. Three of the nine species of lovebirds exhibit sexual differences. The females of these three species tend to have less colorful plumage than males.

Love birds bond strongly to their partners and like to sit in pairs, preening each other's feathers for long periods. Three species are often kept in captivity and can breed readily. The other species are rarer and tend to exist only in the wild. The most endangered lovebird is the black-cheeked, a small green lovebird, with a black face and bright white eye-rings. About ten thousand of these birds live within a few thousand square miles in Southwestern Zambia.

The rosy-faced lovebird of southern Africa has developed an ingenious adaptation to its nesting requirements. It can pack long stems of grass or wood fibers securely under its rump feathers. Then it flies home with its cargo and uses the fibers to line its nest.

Parakeets and Hanging Parrots

Parakeets are prevalent in Africa, India, and Malaysia, sometimes near human settlements, where they feed on ripening grain.

Many of these small birds are green, with head markings of lavender, pink, or orange. They survive in a wide diversity of habitats. For example, the ring-necked parakeet is found in tropical Africa and in the warmer parts of Southeast Asia. The derbyan parakeet lives high in the Tibetan mountains. On the Mascarene Islands, southeast of Madagascar, lives the Mauritius parakeet, now considered one of the rarest birds in the world. Slaty-headed parakeets live the farthest north of any other parrot in the world—thirty-four degrees north latitude in eastern Afghanistan.

The hanging parrots of India and Malaysia are a species of parrots mostly known for their habit of sleeping upside down like bats. Groups of them spend the night hanging inside the leafy canopies of trees. These birds are tiny, and their plumage is mostly green, with blue, red, or orange markings on the head and wings. One species of this type of parrot, the ceylon lorikeet resembles the rosy-faced lovebird in its ability to pack green leaf edges under its rump feathers, to be used to line its nest.

Lories have strong feet and very sharp nails, allowing them to hang sideways or even upside down when feeding. This red lory from the Moluccan Islands has a loud voice, capable of high-pitched screeches.

Two rose-ringed parakeets interlocking beaks. These birds are so common in urban areas of India that they are considered pests.

Following page: A male eclectus. Of all parrots, these birds show the most marked sexual dimorphism. Cocks are green with red underwing coverts. Hens are mostly brilliant red, with a thick band of purple across their bellies.

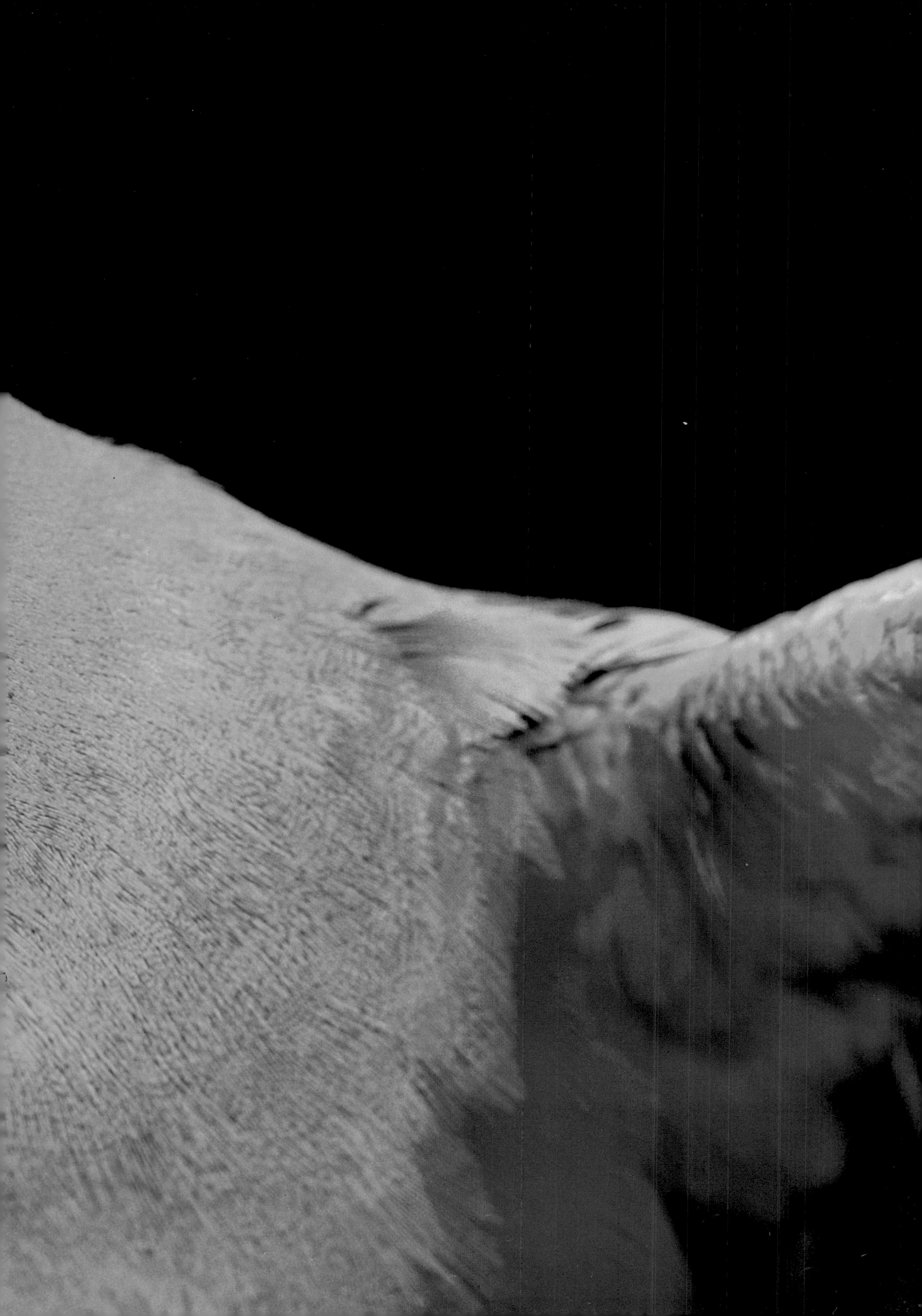

This drab-looking Meyer's parrot displays a vivid blue rump when it takes flight. The contrast provides perfect opportunities for dramatic signaling to its companions. Its wingbutts are bright yellow, useful for intimidating other threatening birds.

AFTERWORD

What do the echo parakeet from Mauritius, the red-tailed black cockatoo from Australia, and the Spix's macaw from Brazil have in common? All are among the dozens of parrot species that are now threatened by extinction. There are international conservation efforts designed to protect these birds, but not all of them are guaranteed to succeed.

This brief survey of the world of parrots has only hinted at the enormous variety and ingenuity of these animals. Unfortunately, their assets have not always worked in their favor. Parrots' relationship to humans has always been problematic. They've been treated as pets, pests, oddities, and even sources of food in some cultures. Their special needs within their natural habitats have not been taken into account until recently. Hopefully, it is not too late to do so. It is becoming more and more evident that the trade of parrots caught in the wild must in many cases be outlawed or severely limited. In too many cases, parrots' habitats need to be preserved or immediately replaced. Their survival must be ensured by supplemental breeding efforts in aviaries. The times call for increased understanding and empathy for this vast family of birds, which includes some of the most beautiful and intelligent animals on the planet.

The purple crested lory of southern Africa can be distinguished from other lories by its long, pointed crest and slightly dark back plumage. It has a loud, squawking call.

A Bourke's parrot from southern or central Australia. With a hatching period of only eighteen days, these birds can produce quite a lot of offspring in one season, which lasts from March to September.

INDEX

Page numbers in **boldface** type indicate photo captions.